THIS BOOK IS THE PROPERTY OF:

STATE_____

PROVINCE_____

COUNTY_____

PARISH_____

SCHOOL DISTRICT_____

OTHER_____

Book No. _____
Enter information
in spaces
to the left as
instructed

ISSUED TO	Year Used	CONDITION	
		ISSUED	RETURNED

PUPILS to whom this textbook is issued must not write on any page or mark any part of it in any way, consumable textbooks excepted.

1. Teachers should see that the pupil's name is clearly written in ink in the spaces above in every book issued.
2. The following terms should be used in recording the condition of the book: New; Good; Fair; Poor; Bad.

Frog Ran

written by Maryann Dobeck
illustrated by Priscilla Burris

SAXON
PUBLISHERS

Frog ran off.

Pig ran, ran, ran.

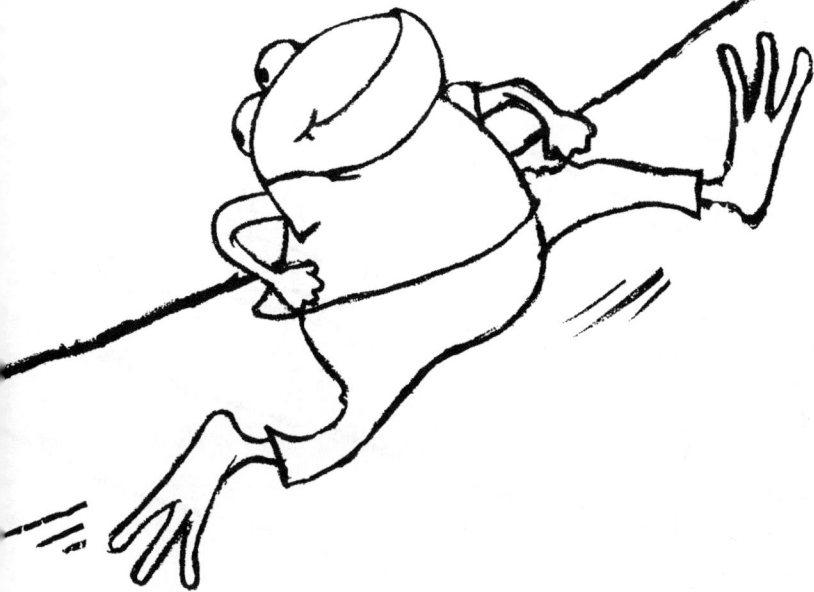

Frog ran to the top.

Frog sat on the hill.

Pig got to the top.

Pig still ran.

Understanding the Story

Questions are to be read aloud by a teacher or parent.

1. What are Frog and Pig doing?

2. Who wins the race?

3. Why didn't Frog win the race?

Answers: 1. Possible answer: having a race 2. Pig 3. Possible answer: because he sat down to rest

Saxon Publishers, Inc.
Editorial: Barbara Place, Julie Webster, Grey Allman, Elisha Mayer
Production: Angela Johnson, Carrie Brown, Cristi Henderson
Brown Publishing Network, Inc.
Editorial: Marie Brown, Gale Clifford, Maryann Dobeck
Art/Design: Trelawney Goodell, Camille Venti, Jillian Gordon
Production: Joseph Hinckley

© Saxon Publishers, Inc., and Lorna Simmons

Printed in China
ISBN-10: 1-56577-950-9
ISBN-13: 978-1-56577-950-1

34 35 36 37 0940 20 19 18 17
4500690836

SAXON
Phonics and Spelling
K

Phonetic Concepts Practiced

r (ran, Frog)

ISBN-10: 1-56577-950-9
ISBN-13: 978-1-56577-950-1

Grade K, Decodable Reader 4
First used in Lesson 51